D1606650

Life Under a STONE

Malcolm Penny

Chicago, Illinois

Published by Raintree, a division of Reed Elsevier, Inc.
Chicago, Illinois
Customer Service 888-363-4266
Visit our website at www.raintreelibrary.com

For information, address the publisher:
Raintree, 100 N. LaSalle, Suite 1200, Chicago, IL 60602

Project Editors: Geoff Barker, Marta Segal Block, Jennifer Mattson, Kathryn Walker
Production Director: Brian Suderski
Illustrated by Dick Twinney
Designed by Rob Norridge
Original design by Ian Winton
Picture research by Rachel Tisdale

Planned and produced by Discovery Books

Library of Congress Cataloging-in-Publication Data:
Penny, Malcolm.
Life under a stone / Malcolm Penny.
p. cm.
Summary: Looks at the environment and organisms that exist under stones.
Includes bibliographical references (p.).
ISBN 0-7398-6806-3 (lib. bdg. : hardcover)-- ISBN 1-4109-0352-4 (pbk.)
1. Soil animals--Juvenile literature. 2. Soil ecology--Juvenile literature.
[1. Soil animals. 2. Soil ecology. 3. Ecology.] I. Title.
QL110.P45 2003
577.5'7--dc21
2003002660

Printed and bound in the United States.
07 06 05 04 03
10 9 8 7 6 5 4 3 2 1

Acknowledgments
The publishers would like to thank the following for permission to reproduce photographs:
Cover: Jane Burton/Bruce Coleman; p.9: Anthony Bannister/Natural History Photographic Agency; p.8: Oxford Scientific Films; p.9: Ernie Janes/Natural History Photographic Agency; p.10: Hans Reinhard/OKAPIA/Oxford Scientific Films; p.11: G.I.Bernard/Natural History Photographic Agency; p.13: Scott Camazine/Oxford Scientific Films; p.14: Tim Shepherd/Oxford Scientific Films; p.15: G.I.Bernard/Natural History Photographic Agency; p.17: G.I.Bernard/Natural History Photographic Agency; p.18: Oxford Scientific Films; p.19: Paulo De Oliveira/Oxford Scientific Films; p.20: Daniel Heuclin/Natural History Photographic Agency; p.21: John Cancalosi/Bruce Coleman; p.22: Stephen Dalton/Natural History Photographic Agency; p.23: Guy Edwardes/Natural History Photographic Agency; p.24: Stephen Krasemann/Natural History Photographic Agency; p.25: Michael Leach/Natural History Photographic Agency; p.26: Kim Taylor/Bruce Coleman; p.27: Zig Leszczynski/AA/Oxford Scientific Films; p.28: Stephen Dalton/Natural History Photographic Agency; p.29:David Boag/Oxford Scientific Films.

Some words are shown in bold, **like this.** You can find out what they mean by looking in the glossary.

Contents

Under a Stone

Shelter from the Elements

Under a stone the world is dark and damp. Dirt, trash, and dust collect there, and rainwater trickles in. It may not sound like a nice place to us, but for some living things, it is perfect. Animals need moisture to survive and the stone protects them from getting dried out by wind and sunshine. The air there stays moist, even if it is dry outside. Food is blown in by the breeze, or washed in by the rain, though some animals who live there leave the stone to go hunting.

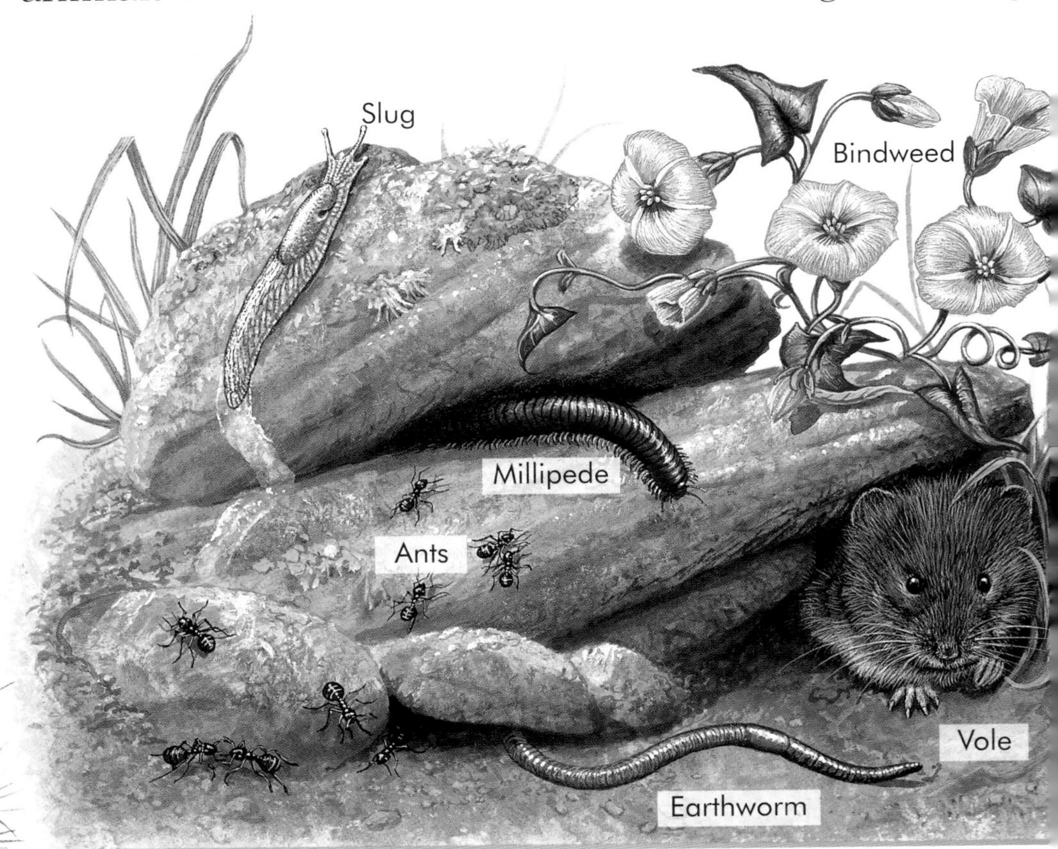

Teeming with Life

Some stones are very large, sheltering huge colonies of animals that are well **adapted** to the conditions there. But even small stones offer the same benefits to the animals that make use of them. When we turn stones over—whether in the wilderness, the park, or in our own backyards—we find many fascinating creatures living in their own secret **microhabitat**.

Guess What?

- Even in deserts, where the daytime air temperature is often more than 95 °F (35 °C), the temperature under a stone stays below 86 °F (30 °C).
- Springtails, tiny insects that often live under stones, are among the few insects that can live in the permanent snows inside the Arctic and Antarctic circles.

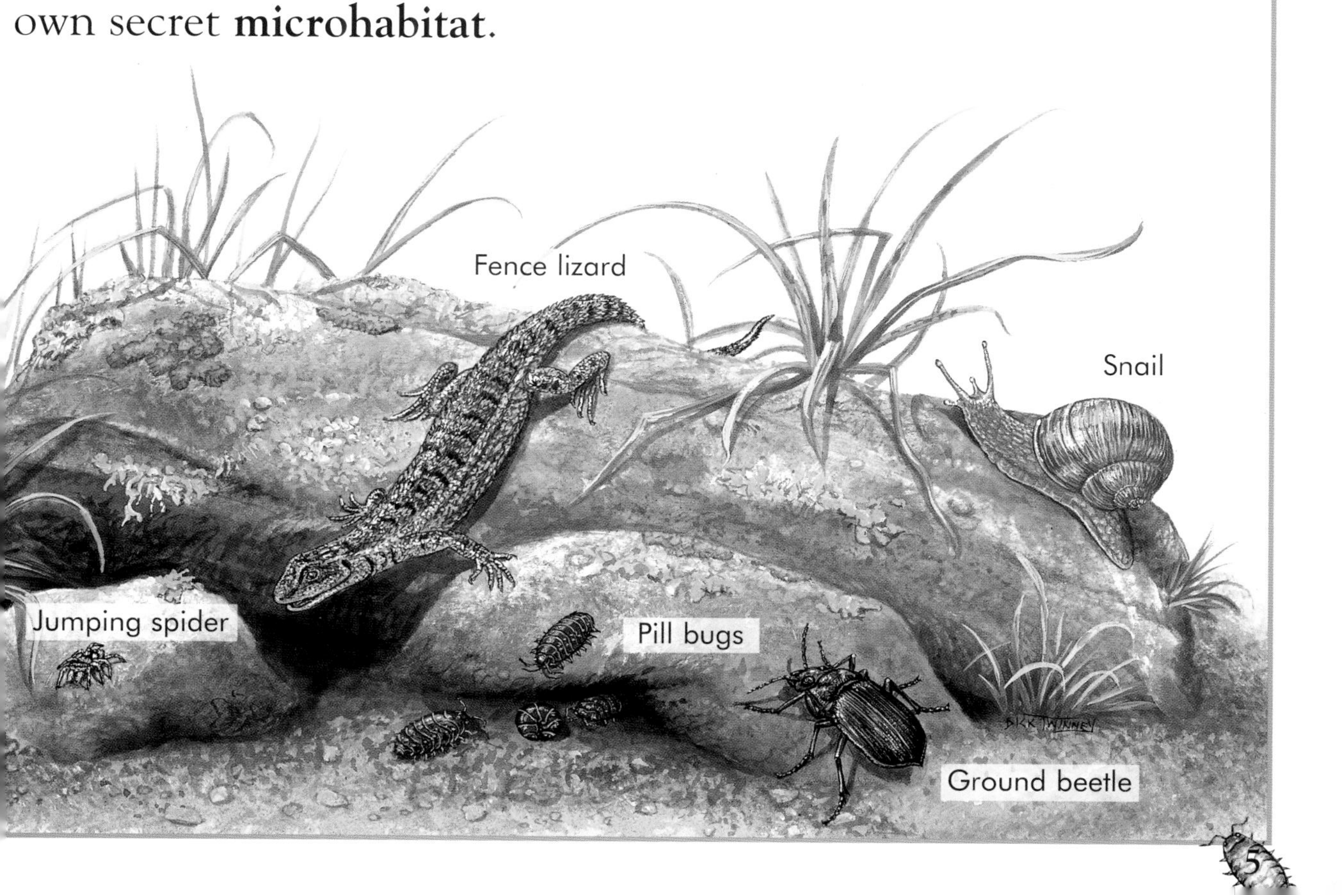

Living in the Dark

Because it is dark under a stone, eyes are not very useful and so stone dwellers are usually blind. Instead, they use the senses of touch, smell, and taste to survive.

In the tight space beneath a stone, it helps to be small. Tiny insects like ants and springtails can move easily under the low roof. Larger insects like beetles need hard bodies to push under the stone and strong legs for **burrowing** in the soil. The hard front wings of beetles, called **wing cases,** close around their bodies like armor, and many of them have powerful legs.

Springtail

Pill bug

Flat animals are also well suited to life under a stone. Centipedes scuttle around, hunting insects and other small animals, and so do flat-bodied roaches with their tough wing cases.

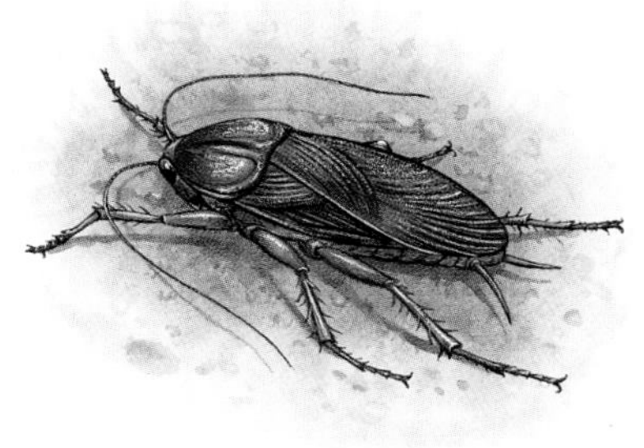

Roach

Invaders

Some animals that do not live under stones visit them to hunt other animals for food. **Predators** like ground beetles and rove beetles often raid under stones, and larger animals like anteaters and raccoons can destroy the community completely by turning the stone over in their search for **prey.**

Guess What?

Huge roaches that live under stones in Madagascar can be 6 in (15 cm) long. When picked up, they make an alarming squeaking noise that can startle would-be predators.

The **larvae** of ground beetles eat like spiders. They inject juices from their own bodies into their prey (such as a slug or snail). This turns the insides of the prey into a mushy soup that they then suck dry.

This South African ground beetle has huge jaws, and golden whiskers decorating the middle section of its body.

Slitherers and Wrigglers

Slugs and Snails

The damp darkness under a stone is perfect for animals that can slide and slither. Slugs and snails lay their eggs along the edges of stones so that the newly hatched young can crawl underneath. The stone protects them from their enemies, and from the sun, which would harm them by drying their skin. When a snail grows too big, it cannot live under a stone anymore because its shell will not fit. But slugs, because they have no shell, can continue to take shelter under stones throughout their lives.

Tiny snails hatch from eggs laid close to the edge of a stone.

Leaving a Trail

Both slugs and snails leave a trail behind them as they crawl along. The trail is made of **mucus**, a slimy substance produced by **glands** under their body. The mucus makes it easier for them to slip over the ground, but it is also sticky. This is how slugs and snails can climb on walls and even up windows.

Adult snails are protected by their shells, but they still have enemies that eat them—including some people!

Earthworms, too, produce mucus, to help them to slide along in their **burrows**. Unlike slugs and snails, which are smooth underneath, earthworms have tiny bristles on their undersides to help them grip the soil.

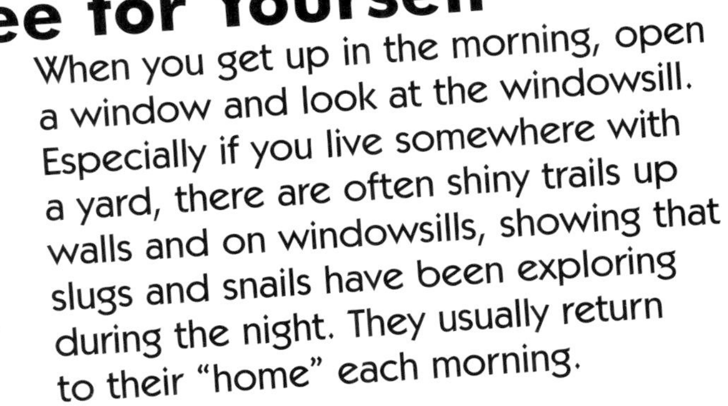

See for Yourself

When you get up in the morning, open a window and look at the windowsill. Especially if you live somewhere with a yard, there are often shiny trails up walls and on windowsills, showing that slugs and snails have been exploring during the night. They usually return to their "home" each morning.

Wriggly Hunters

Most of the animals that live under stones are **invertebrates**, which means animals without backbones. Some people call them creepy-crawlies. Among the creepiest and certainly the crawliest are centipedes.

A centipede has a large head with powerful jaws that it uses to inject poison into the animals it hunts.

Centipedes are ideally suited for life under a stone. They are flat, and they move by wriggling and scuttling along with one pair of legs on each segment of their long bodies. They hunt small insects, including springtails, and also the eggs and newly hatched young of slugs and snails. Centipedes hunt outside as well, crawling from under the stone at night and returning to hide there during the day.

Leaf Chewers

Millipedes are not flat like centipedes, but rounded, tubular animals. Neither are they hunters—they are **vegetarians.** They like to live under stones, chewing up dead leaves and other plant material, and **burrowing** into the soil. By doing this, they mix the plant material into the soil, making it richer for the other animals that live there.

Guess What?

- The biggest centipede in the world is the giant scolopender in the rain forests of Central and South America. It grows to over 10 in (25 cm) long.
- Millipedes have two pairs of legs per segment, while centipedes have only one pair per segment.
- Some tropical forest millipedes can glow in the dark!

When it is in danger, a millipede defends itself by curling up into a tight spiral, making it harder to attack.

The Burrowers

Underground Experts

Many beetles are experts at **burrowing**. Ground beetles, with hard **wing cases** as long as their bodies, can burrow into the soil under a stone. Rove beetles have short wing cases and therefore more bendy bodies, so they can crawl into cracks where ground beetles cannot go. Both kinds are perfectly at home under stones.

Ground beetles have flattened bodies and long wing cases.

The smallest beetles can find all they need to eat under the stone where they live. But larger beetles must go out hunting, usually at night, and go back under the stone to hide and rest during the day.

A rove beetle's short wing cases cover less than half of its body.

Gravediggers

One kind of beetle, the burying beetle, comes out from under the stone to look for food, but not to hunt. Burying beetles live up to their name by burying already dead animals, such as mice or baby birds. They scrape away the soil from under the **carcass** until it sinks out of sight.

When the adults have fed from the corpse, the female lays eggs on it. The **larvae** eat the dead body when they hatch. After laying the eggs, the beetles hide under a stone until the next night's work begins.

Guess What?

- Most insects live for less than one year, but one burying beetle that accidentally got trapped inside a block of concrete was found still alive when the block was broken open 16 years later.
- The fastest moving beetle is the tiger beetle, a type of ground beetle that can run at 2 ft (60 cm) per second—a little over one mile per hour.

Burying beetles are sometimes called sexton beetles, because sexton is an old name for gravedigger. These ones are burying a dead chipmunk.

Not-so-good Burrowers

Earwigs look like beetles, but they do not have as strong legs and their bodies are not so well armored. Rather than **burrow,** they creep into cracks between stones or into gaps underneath them. They are more at home in damp leaf litter, but if they cannot find that, a shelter under a stone will do. They feed on living and dead plants as well as small insects.

A mother earwig surrounded by her young under a safe, damp stone.

The **pincers** at the end of an earwig's body are a weapon for attacking **prey**, but the earwig uses them mainly to defend itself. It tries to bite anything that attempts to pick it up—including people.

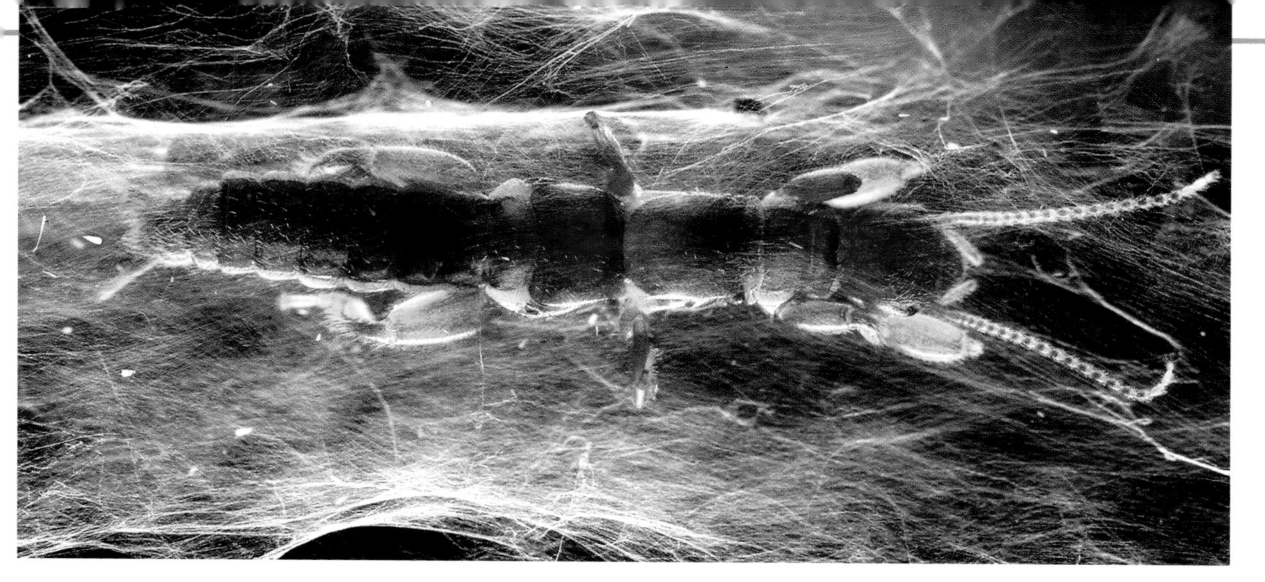

A tiny web spinner in its silky tube in Trinidad, West Indies.

Web Spinners

Another group of insects that live under stones look like tiny earwigs with wings. Their pale, golden bodies are just one-fifth of an inch (half a centimeter) long. They are web spinners, and they live mainly in the **tropics,** though some are found in the southern United States.

Web spinners live in silk-lined tunnels running under stones or in the soil. The silk comes from **glands** in the web spinners' front legs. They feed mainly on dead plants, which they chew up with tiny, sharp jaws.

Guess What?

In past times when people slept on straw mattresses, earwigs would sometimes crawl into people's ears. The word earwig came from the phrase "ear wriggler."

A colony of web spinners is made up of a mother, a father, and their young—sometimes as many as a hundred of them.

Tunneling Ants

Enormous numbers of ants live under stones. We can tell that they are there by the piles of loose dirt that they bring to the surface as they dig out their **burrows**. They carry the sand or soil particles out from under the stone and leave them near the edge. The nests that they build contain many tunnels, which widen out in places to form chambers.

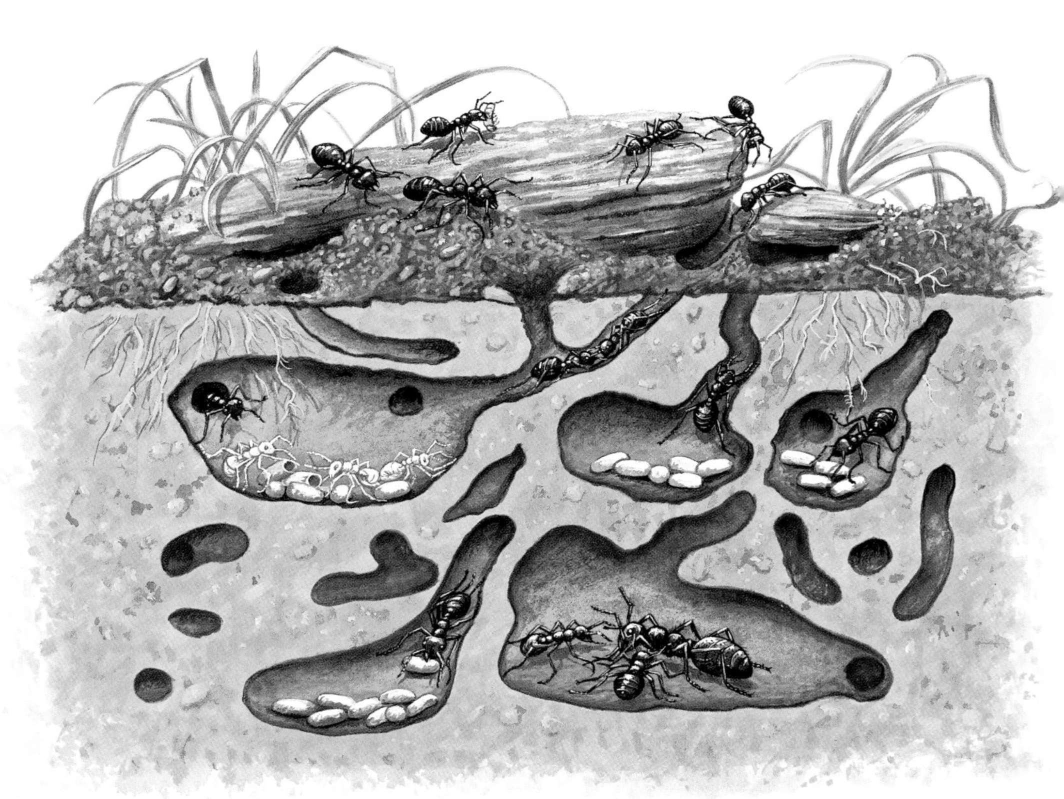

Some chambers inside the ants' nest are used for storing food, and others are for raising young ants. One chamber is used by the queen ant (seen here at the bottom right of the picture), the leader of the colony.

Complex Colonies

In a working colony, there are two and sometimes three different kinds of ants, all of them female. There is one queen who lays all the eggs, thousands of workers who look after the eggs and **larvae**. Some ant colonies also have a few soldiers who guard the workers while they are out gathering food.

In late summer, a fourth kind of ant appears—the flying adults. Soon the winged males and females leave the nest together and fly into the air. After a pair has mated, they drop to the ground. The males die and the female's wings either fall off or are broken off. Then the female crawls away to find a suitable place to start a new colony. Under a stone is a good place to begin.

Winged ants leaving their colony in the warm, dry weather of August.

Eight-legged Dwellers

Sharp-eyed Hunters

Many spiders build webs to catch flying insects, but others trap their food in different ways. Jumping spiders carefully creep up on their **prey** and then suddenly pounce on it. Although you often find jumping spiders living under stones, they do not hunt there because it is too dark. They have to be able to see their prey to catch it.

A jumping spider has two large, forward-facing eyes and six smaller ones, for pinpointing its prey.

Another stone-dwelling hunter, the woodlouse spider, has long jaws that it uses to hunt woodlice, also known as pill bugs.

Booby Trap

The tube-web spider lies in wait for its prey under stones. It makes its lair by lining a hole with silk. Leading out from the lair are other strands of silk that wiggle if insects touch them. If the spider feels one of the strands move, it rushes out to grab the insect. The spider bites its prey, then drags it back into the lair to eat. When it has finished eating, it throws out the dry, empty body of the insect.

See for Yourself

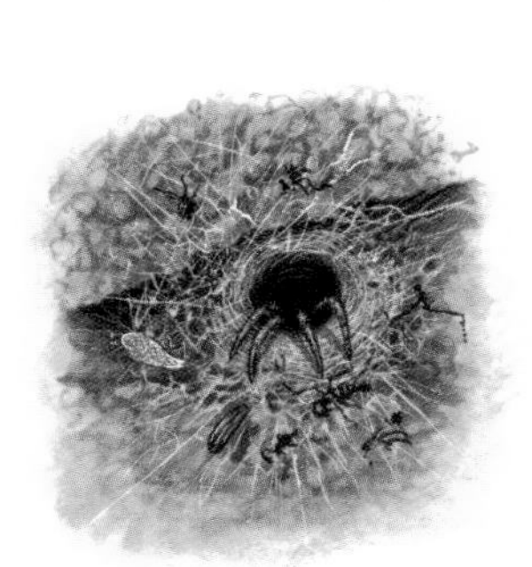

Tube-web spider lairs are not hard to find. A small hole, under a stone or in the corner of a wall, has a gray funnel at the entrance, with fine lines leading out from it. There are dry, dead insects around the entrance. If you look inside very carefully, you might see the spider's front legs.

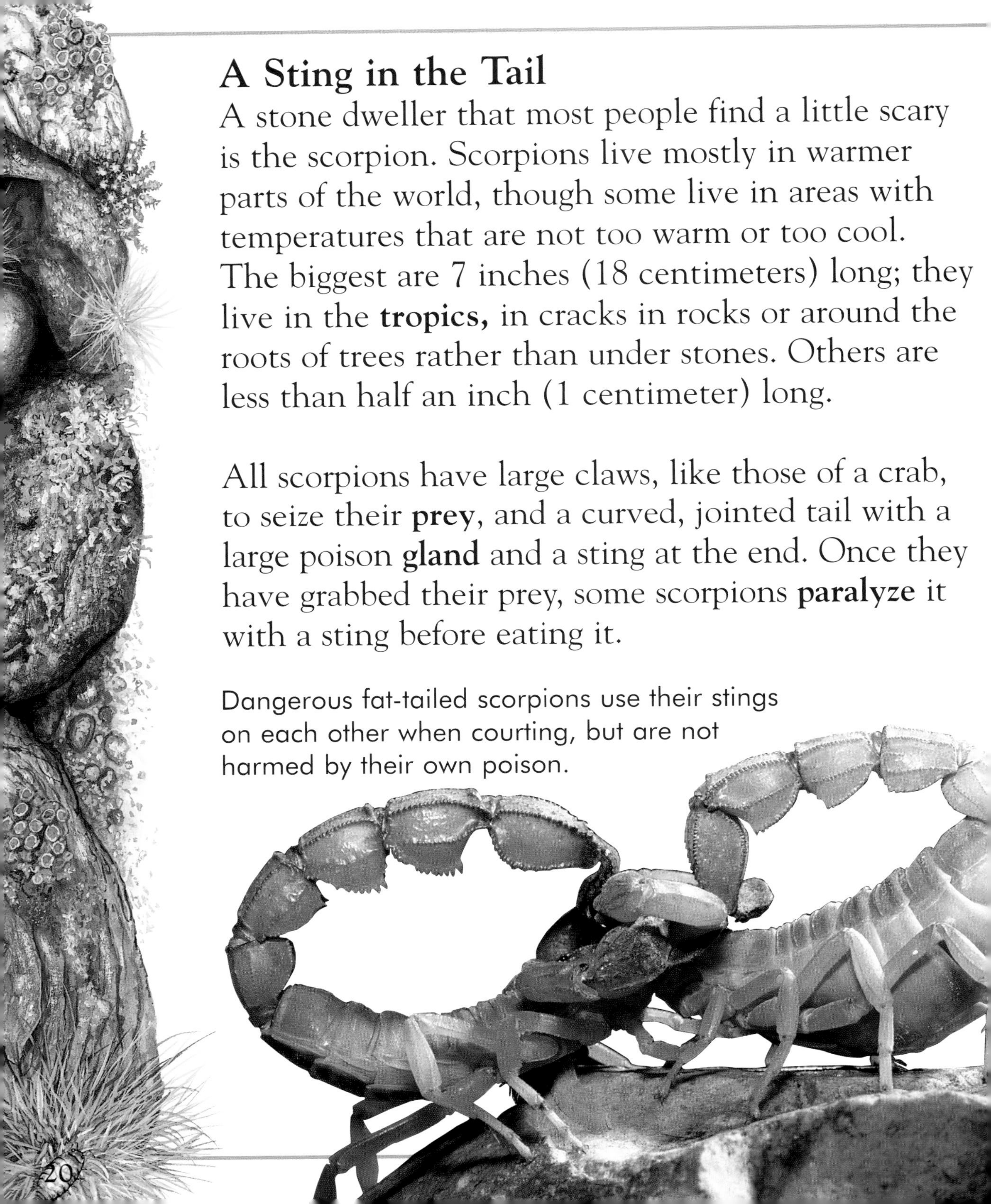

A Sting in the Tail

A stone dweller that most people find a little scary is the scorpion. Scorpions live mostly in warmer parts of the world, though some live in areas with temperatures that are not too warm or too cool. The biggest are 7 inches (18 centimeters) long; they live in the **tropics,** in cracks in rocks or around the roots of trees rather than under stones. Others are less than half an inch (1 centimeter) long.

All scorpions have large claws, like those of a crab, to seize their **prey**, and a curved, jointed tail with a large poison **gland** and a sting at the end. Once they have grabbed their prey, some scorpions **paralyze** it with a sting before eating it.

Dangerous fat-tailed scorpions use their stings on each other when courting, but are not harmed by their own poison.

A mother scorpion with young on her back is one creature you might prefer not to find under a stone.

Caring Mothers

Scorpions are very careful mothers. The mother does not lay her eggs, but keeps them inside her body in a special pouch until they hatch. The young crawl out and immediately climb up on to her back, where they stay until they are about two weeks old.

Guess What?

Scorpions have eight legs and are **arachnids,** just like spiders.

The world's most dangerous scorpion is the fat-tailed scorpion found in Algeria, North Africa. Its sting can kill a human in four hours and a dog in seven minutes.

Most scorpions do not have venom that is very poisonous to humans. Usually a scorpion sting causes no more harm than a wasp sting.

Larger Residents

Sun-loving Hunters

Lizards (below right) hide under stones and come out to look for **prey**, usually in the daytime. They cannot make their own body heat, so they need to lie in the sun to warm up. But lizards also need to escape from the sun during the warmest time of the day or they will get too hot. Under a stone is a good, cool place to take shelter.

Vegetarian Voles

Some of the **burrows** that can be found under stones are made by voles. These tiny **vegetarians** do not find their meals under the stone. Instead, they come out at night to hunt for seeds or nuts, or sometimes to dig up and eat the roots of plants.

A mother vole brings her young out of their burrow.

Talented Tunnelers

There is one animal who often hunts under stones but very rarely comes out at all. The mole lives almost entirely underground, tunneling through the soil to find earthworms.

Guess What?

- The tail of some lizards will drop right off if it is pulled or grabbed by an enemy, allowing the lizard to escape. It will grow a new tail later.
- Many people think moles are blind, but they can tell light from darkness. In very dry weather, they use their vision to find streams where they can drink.

As it digs, the mole needs to push the soil from its tunnel to the surface, but where its burrow passes under a large stone it cannot do this. When a stone blocks the way, the mole makes a side tunnel and pushes the soil out through there instead. If you find a flat stone with a molehill beside it and lift the stone, you might see the mole's tunnels underneath. Each one is no more than about 1 inch (3 centimeters) across.

A mole pops its nose out for a rare glimpse of daylight, between its shovel-like feet.

Plant Life

The Need for Light

Almost all plants need sunlight to grow. They trap the energy from the sunlight with a chemical called **chlorophyll**. This energy is used to combine water and carbon dioxide from the air into sugars that the plant can store and use. This process is called **photosynthesis.** Because they rely on light to survive, no plants can live completely hidden under a stone.

The Indian pipe isone of the few plants that can live without sunshine. It does not need to use sunlight for energy because it steals food from the roots of other plants.

Hidden Glory

However, stones may protect some plants. Bindweed is sometimes called "morning glory" because it closes its flowers at night, and opens them again at sunrise. The sugars that the bindweed has made are stored in thick roots underground, and its long stems wind around other plants as it climbs to the sunlight. If its roots are under a stone, the plant uses its stored energy to grow pale stems in the darkness that reach around the stone. It produces leaves with chlorophyll only when the stems reach the sunlight.

The trumpet-shaped flowers of bindweed open up in the morning.

See for Yourself

Place a stone on a patch of green grass. After about a week, remove the stone and you will see that the grass under it has turned yellow. This is because chlorophyll, the chemical that traps energy from sunlight and makes plants look green, breaks down if deprived of light for more than a few days. Without chlorophyll, plants cannot live. If the stone is left on the grass, after a few more weeks the grass will die.

Under Desert Stones

Flat stones

Flat stones are a very important **microhabitat** in deserts, because they provide shelter from the hot, dry air. In the Middle East, Africa, and parts of North and South America, these stones may be the daytime home of fierce **predators** like sun spiders (also called camel spiders), that come out to hunt only in the cool of the night.

This sun spider from eastern Africa is eating a grasshopper and ignoring the ants that are trying to bite it.

The sun spider is not really a spider, but is an **arachnid**, so it is related to spiders in the same way as a scorpion. Sun spiders sometimes raid the nests of hornets, which **burrow** deep into the ground, often starting the burrow under a flat stone that shelters the entrance like a porch.

Shelled Shelterer

In eastern Africa there is a tortoise whose shell is so flat that it can live in shallow cracks under rocks. But even there it is not safe. The chanting goshawk is a **bird of prey** with very long, flexible legs that can reach under rocks to pull the tortoises out.

Guess What?

Hornets buzz loudly to warn intruders that are too near their nest, but although they have powerful stings they rarely have to use them: other animals, including humans, are usually too scared by the noise to come close.

A pancake tortoise squeezes under a rock to hide from its enemies.

Stones Under Threat

Turning a World Upside Down

The greatest threat to animals that live under stones is that the stone might be turned over. Sometimes **predators** flip stones while they are looking for **prey**, but often it happens because people do not realize the damage they can do by driving through wild areas in off-road vehicles, or even by accidentally kicking stones. A community that has grown up over years can be destroyed in a moment.

Sometimes microhabitats are destroyed because of human curiosity.

Dirty Water

If stones lie near parking lots or along roads, the animals that live under and around them may be harmed by **polluted** water. This is because water washes across the pavement where it picks up oil and other dangerous chemicals, and then ends up under the stones where it will damage the **microhabitat.**

A common toad hides under a stone watching for prey to pass by.

Leave Stones Alone

In people's backyards, stones are just as important to small creatures as they are out in the wilderness. But a lot of people do not want their backyards to look messy, so they sweep away the dust that gathers in corners and arrange stones in neat lines and piles. Their backyards might be clean, but a more natural yard would have a lot more life in it. Sometimes it is better to leave one or two stones unturned.

Guess What?

Centipedes and ground beetles are gardeners' friends. They take shelter under backyard stones, killing slugs and young snails that munch on garden plants.

Some parking lots are designed so that water soaks through the pavement without running off. The holes that drain the water allow small animals to burrow into the soil below. This creates a microhabitat under the parking lot, as if the pavement were one huge, flat stone.

Glossary

adapted changed slowly over time to be able to survive in a particular place

arachnid invertebrate with eight legs and no antennae, whose body is divided into two parts. Scorpions, spiders, and ticks are all arachnids.

bird of prey bird that hunts animals to eat

burrow dig a hole or tunnel in the ground. A hole or tunnel used as an animal's home is called a burrow.

carcass body of a dead animal

chlorophyll green chemical in plants that captures energy from sunlight

gland part inside the body that produces substances an animal needs to survive

invertebrate animal without a backbone

larva (more than one are called larvae) the wormlike stage in a young insect's life, when it looks nothing like its parents

microhabitat special environment, such as under a stone, where particular animals or plants live

mucus any sticky, slippery substance made by an animal's body. A snail or slug oozes mucus that helps it slide along the ground.

paralyze to make something unable to move

photosynthesis process by which plants take in sunlight and water and change them into energy with the help of chlorophyll

pincers claws for pinching, such as those on an earwig

polluted spoiled, or made impure (especially with artificial waste)

predator animal that hunts other animals for food

prey animal that is hunted by other animals for food

tropics very warm area on Earth located around the equator

vegetarian animal that does not eat meat

wing cases hard wings on certain insects, such as beetles, that cover and protect the more fragile wings used for flying

Further Reading

Fredericks, Anthony and Jennifer DiRubbio. *Under One Rock: Bugs, Slugs, and Other Ughs*. Nevada City, Calif.: Dawn Publications, 2001.

Green, Jen. *Under a Stone*. New York: Crabtree, 1999.

Greenaway, Theresa. *Minipets: Beetles*. Chicago: Raintree, 2000.

Greenaway, Theresa. *Minipets: Centipedes and Millipedes*. Chicago: Raintree, 2000.

Greenaway, Theresa. *Minipets: Slugs and Snails*. Chicago: Raintree, 1999.

Greenaway, Theresa. *Minipets: Spiders*. Chicago, Raintree, 1999.

Murray, Peter. *Beetles*. Eden Prairie, Minn.: Child's World, 2003.

Richardson, Adele D. *Scorpions*. Minnetona, Minn.: Capstone Press, 2002.

Index